ODDITIES OF NATURE

By
BUD JONES

© Copyright 2013, Bud Jones

All Rights Reserved.

No part of this book may be reproduced, stored in a
retrieval system, or transmitted by any means,
electronic, mechanical, photocopying, recording,
or otherwise, without written permission
from the author.

ISBN: 978-1-304-47960-0

Dedicated to

Dan Spaulding

A botanist without peer and a friend

Foreword

Most of my life I have been interested in nature. At the age of ten years I remember finding a dead blue jay, and I asked my mother if she thought that, if I covered it with glue, would it be preserved. I will never forget her kind answer.

"No, Son, I don't believe that would do any good."

This lifelong interest in things natural eventually led to my becoming a professional taxidermist, exposing me to animals from North America, South America, India, Mexico, Africa, Canada, and other far-flung countries. In one of my books, <u>Some Wild Animals I have Known</u>, I told about many of the animals that I have come in contact with. However, I wanted to do more. I wished to write a book on common plants and animals, common but still not too familiar to most people, hence this publication, <u>Oddities of Nature</u>. It is a book about things that go on around us every day, yet most people fail to see them, and those that do see them have no understanding of what they are looking at.

This is not a scientific treatise—far from it. Hopefully it is a book that everyday lay people can read and truly understand without having to scratch their heads and wonder. The scientific names are used so that if any person wishes to pursue matters further, then the animal or plant can be tracked down more easily.

I hope you, dear reader, will enjoy this book as much as I have in its creation.

Bud Jones July 2010

Table of Contents

Chapter 1 – The Spittle Bug 1

Chapter 2 – The Trapdoor Spider 3

Chapter 3 – The Ant-aphid Relationship 5

Chapter 4 – The Box Turtle 7

Chapter 5 – Where Do Butterflies Spend The Night? 9

Chapter 6 – Can A Snake Climb A Tree? 11

Chapter 7 – The Fence Lizard 13

Chapter 8 – The Dung Beetle 15

Chapter 9 – The Dragonfly 17

Chapter 10 – The Oak Gall 19

Chapter 11 – Abbott's Sphinx Moth 21

Chapter 12 – The Flying Squirrel 23

Chapter 13 – Jewelweed 25

Chapter 14 – The Freshwater Jellyfish 27

Chapter 15 – The Buzzards Are Back 29

Chapter 16 – The Ant Lion (Doodle Bug) 31

Chapter 17 – The Glass Lizard 33

Chapter 18 – The Narrow-Mouthed Toad 35

Chapter 19 – The Velvet Ant 37

Chapter 20 – Lichen 39

Chapter 21 – The Tent Caterpillar 41

Chapter 22 – The Tiger Swallowtail 43

Chapter 23 – The Caddisfly Larva 45

Chapter 24 – The Amazing Swallowing Apparatus of Snakes 47

Chapter 25 – The Mushroom That Grows In Manure 49

Chapter 26 – The Short-Tailed Shrew 51

Chapter 27 – The Hercules Beetle 53

Chapter 28 – The Chuck-Will's-Widow 55

Chapter 29 – What About The Sex Life Of Snakes. How Do They Breed? 57

Chapter 30 – The Praying Mantis 59

Chapter 31 – Pinesap – A Strange Plant 61

Chapter 32 – The Flat-headed Worm 63

Chapter 33 – The Guidance System of the Canebrake Rattlesnake) 65

Chapter 34 – The Cicada – A Life Spent Underground 67

Chapter 35 – Regeneration – The Five-Lined Skink 69

Chapter 36 – The Mud Dauber 71

Chapter 37 – The Armadillo 73

Chapter 38 – The Ichneumon Wasp 75

Chapter 39 – The Braconid Wasp 77

Chapter 40 – Another Species of Ant Lion 79

Chapter 41 – The Vole 81

ODDITIES OF NATURE

Name: Spittle Bug larva
Date Received: May 2010
Address: Tallapoosa, GA
Family: Cercopidae

Chapter 1
The Spittle Bug
(Family C*ercopidae*)
Class Insecta

Many years ago as I walked across a weed covered field my attention was captured by a big wad of spit on the stem of a goldenrod plant. I shook my head, for I knew that no one but me had entered that field on that particular day. The "spit" was bubbly looking and indeed appeared strange to me. Finally I took a twig and probed into it. Imagine my surprise when I uncovered a tiny "bug" inside that accumulation of "spit."

That was my first introduction to the spittle bug, one of Mother Nature's anomalies. Actually this strange creature is the nymph of an insect which later turns into an adult insect called a leaf hopper, or frog hopper. The nymph itself is responsible for the frothy spit. It does this by sucking juice from the host plant, and as that juice passes through and out the alimentary tract, the nymph swishes its tail end back and forth to create the frothy spit. This mass of liquid not only keeps the nymph's skin damp but it also protects it from the sharp-eyed birds that prey upon it.

Finally, after about six weeks, the nymph turns into an adult, eggs are laid on the host plant, and the spittle bug life cycle starts all over again.

Bud Jones

Chapter 2
The Trapdoor Spider
(Cyclocosmia truncata)

I was amazed when I saw my first trapdoor spider. Over an inch long, it was a shiny black color with an odd looking, flat-shaped rear end. I knew exactly what it was from the many pictures I had seen. Still, I was not prepared to find one in my own back yard. I searched a long time for its nest but never found it, for it is well camouflaged.

Using rough projections on its jaw edges this amazing arachnid (an animal with 8 legs) digs a tube like burrow. The whole inside of this tube is lined with silk. The trap door itself has a hinge, and is lined with extra silk so that it fits snugly in the rim of the tube. The top of the trap door is made of silk and debris from the ground and is nearly impossible to see.

The trapdoor spider lives in its tube, staying right under the trap door. When it feels the vibrations of a close by insect, it rushes out the trap door to catch it, which could be a centipede, an insect, or even another spider. The captured prey is taken by the spider into the tube nest and devoured at leisure.

Except to capture prey, the female spider hardly ever leaves her tube nest. The males, however, often wander outside, looking for a female with which to mate.

Some scientists suggest that the spider uses its flat ended abdomen to tamp down the silk that lines the burrow itself, which can be several inches deep.

There are numerous species of trapdoor spiders that are found in the Southern United States.

Bud Jones

Chapter 3
The Ant-aphid Relationship
Ant (Hymenoptera) Aphid (Aphididae)

Once I was walking across a pasture and passed by an alder bush. As I looked at the bush I noticed a white substance on one of the limbs. I first thought it was lichen but upon closer examination it looked more like a piece of cotton. Then I saw ants crawling over it and I grew suspicious. I used my finger to rub the white stuff and my suspicions were right. I saw movement, and immediately realized that I had found a plant aphid colony and the ants were taking care of them.

This is one of the greatest examples in nature of a symbiotic relationship – that is – two animals living together for the mutual benefit of both. Aphids are tiny insects that live on plants. The female lays her eggs on the plant itself and the eggs hatch into tiny nymphs. The ones I found were called alder aphids, living strictly on alder bushes (Alnus serrulata). They form a colony and suck the juices from the plant stems. As the juices pass through and out of their bodies, it is excreted as a sugary sweet juice called honeydew, and the ants love it. As they crawl over the colony of aphids they not only dine on those sweet juices but they protect the aphids from other insects that might try to prey on them. So, the relationship is mutual. The aphids furnish the honeydew and the ants give protection to the colony.

To me this relationship is truly a miracle of Mother Nature.

Chapter 4
The Box Turtle
(Terrapene carolina)

To me, box turtles are quaint little animals. They are so quiet and unassuming that it is a wonder that anyone ever pays any attention to them. I have had them for pets, and have watched them cross a road, tucking their heads back in their hard shells as a car passed over them.

Box turtles vary a lot in color but there is usually a good amount of yellow in most of them. The carapace, or top shell, fits closely with the plastron, or bottom shell, so that when frightened the turtles can close their shells so tightly that it is hard to separate them, even when prying with a knife blade.

Male box turtles usually have red eyes, and the bottom shell has a noticeable depression, which enables it to stay on the back end of a female when breeding. The female has yellowish colored eyes and lacks the depression.

When closed in its shell, as mentioned above, the turtle is fairly well protected. Once I saw a dog pick up one, mouth it around a few times, and then drop it, with no apparent damage. A turtle can, of course, be killed easily by a car or a heavy blow, but generally the hard shell serves well when enemies are around.

Box turtles eat a variety of foods such as mushrooms, plant material, bugs, snails, slugs, lizards and salamanders, and even carrion. In winter they dig into the soft dirt or debris, cover themselves, and hibernate.

They can live a long time. I once knew a man who carved his initials on the top shell of a box turtle. Forty years later he found the same turtle on his property within a hundred yards of where he first found it.

Bud Jones

Chapter 5
Where Do Butterflies Spend The Night?
(Order Lepidoptera)

As a boy I collected butterflies, and to this day I have not lost the thrill of chasing natures bright flowers. It has been a lifelong hobby, year after year, as long as I can remember. During all that time I used to wonder what happened to butterflies when it rained, or where did they go when the dark curtain of night closed around them. In all my nights of camping, hunting, and similar activities I had never seen a butterfly resting for the night. On September 27, 2008 that question, so long unanswered, was finally solved.

I had been out on a squirrel hunt and as the shadows began to lengthen I started back to my car. It was getting darker when, about a hundred yards from my car a monarch butterfly flew right across my intended path and lit on the leaf of a persimmon tree. Then it crawled on the underside of the leaf, clinging upside down, and settled in for the long night ahead. I rushed to my car, retrieved my camera, and got a picture of a happening in nature that had perplexed me for so long. I was happy. I had seen a butterfly go to "roost." It was secure under the wide persimmon leaf, and barring any accidents like getting caught by some predator or being dislodged in a severe rain storm, it would yet live another day. Then it could continue its leisurely journey down to Mexico.

Chapter 6
Can A Snake Climb A Tree?
The Gray Rat Snake
(Elaphe obsoleta)

Once I was talking to a friend about snakes and, as snake stories usually go, they can often lead to some wild tales. My friend had seen "some type of snake" in a hickory tree, resting at ease on the branches, and he was astounded that a snake could be in a tree.

Many people do not know that snakes can climb trees. I have seen rat snakes do just that, and twice I have seen a gray rat snake climb a brick wall. My friend was flabbergasted and only believed me when I showed him the picture on the facing page.

It is a photo of a gray rat snake that I found in my woodpile on a very cool October day. The snake had evidently decided to hibernate there for the rest of the winter, for it was curled up and seemingly resting comfortably. I picked it up and handled it awhile and it seemed to "come alive" again. When I finally turned it loose it went over to a mockernut hickory tree in the yard and climbed to the safety of its limbs.

Snakes use the belly scales on their body to grasp rough edges of trees, rock or brick walls, etc., to climb quite freely. I have never seen a heavy-bodied snake such as a rattlesnake, copperhead, etc., do this. I believe that they are not agile enough, yet they can climb to some degree.

Bud Jones

Chapter 7
The Fence Lizard
(Sceloporus undulatus)

The eastern fence lizard has always reminded me of a horned toad. It has rough, scaly skin, is a drab gray-brown color, and the strongly keeled scales give it a "spiny" appearance. On an old fence post or tree bark the lizard can just about disappear, so well does its color blend in with its surroundings.

The adult male lizard has bright blue markings on the throat and on each side of the belly. These markings are heavily outlined in black. The female lacks these markings.

The fence lizard likes dry, open areas that have plenty of trees, stumps, and similar places where it can climb and hide in order to escape its enemies. As a boy, whenever I saw a fence lizard, I would often try to catch it. Like a squirrel it would stay on the opposite side of the tree or fence post, making it most difficult to catch. It would peep around the tree, and as I moved to catch it, a quick scamper around the other side made it impossible to lay a hand on it. I have, however, caught one, only to have a squirming lizard tail in my hand while the wily reptile made his getaway sans his tail.

Mating occurs in the spring, and when egg-laying time comes the female lays about 10 eggs, white and leathery looking, in loose soil or in the loose soil of a rotted log. The female covers the eggs with debris then leaves them, providing no parental care. Once hatched, the baby fence lizard is on his own.

Fence lizards are active only in the daytime. During the spring and summer they can often be seen basking in the sun. When cold weather threatens they will hibernate in wood piles, hollow logs, or similar places.

They are very beneficial animals, eating a large number of insects, spiders, caterpillars, and other invertebrates.

Today they are scarcely seen.

Chapter 8
The Dung Beetle
(Canthon laevis)

Breakfast anyone? Sausage? Scrambled eggs? Bacon or biscuits? No, none of that? Well, what about a nice ball of horse manure? How does that suit your appetite?

As strange as it may seem the dung beetle depends on manure to survive. The female will find a fresh cow or horse pile and proceed to roll some of it into a ball. By doing this the beetle is about to

move the rather large ball (the size of a dime or bigger) to a cool, moist private place where she later will lay her egg. The female does this moving by crawling forward and then pulling the ball behind, using her back legs. The dung balls are then placed in a depression dug by the female, where she lays one egg on the ball. Once hatched, the larva can dine on the manure ball with no competition.

Once the larva has exhausted its food supply it enters into the pupal stage and eventually turns into an adult beetle. Thus the life cycle starts all over again.

Bud Jones

Chapter 9
The Dragonfly
(Odonata)

I have never ceased to be amazed at the flight antics of dragonflies. Over a muddy pool, a creek, or a damp section of earth they dart, twist, turn- about, or fly in a straight line, moving quickly. They might even hover like a helicopter over a certain spot if they choose to do so. Surely they are masters of the air. Sometimes their wings are clear, sometimes marked, and with their large heads and slim, segmented bodies they are truly remarkable insects.

Dragonflies are often called mosquito hawks or snake doctors. The last name comes from an old belief that they attended to sick snakes, which, of course, is a myth. The first name, however, rings true, for as they zip through the air their long legs are held in a basketlike fashion. This enables them to snatch mosquitoes and other flying insects out of their midair flight.

Female dragonflies lay their eggs in the water. Some species lay them on plants. When hatched the nymphs live in water, often scooting around on the creek bottom with quick jerky movements, much like a crayfish. They have a rather large mouth and eat the tiny water animals that live in the same habitat.

The average time spent in the water is about a year. When the time comes to leave the water the nymph crawls out on the bank, fastens itself to a stick or other object, and then splits down the back. Presently, like a butterfly, an adult dragonfly emerges, the wings dry for a short time, and soon it flies off, beginning another life cycle for this amazing insect.

Chapter 10
The Oak Gall
Caused By A Gall Wasp (Andricus seminator)

Once, while on one of my daily walks, I came upon a white oak sapling whose leaves had just formed. Scattered throughout the tree were round, golf-ball shaped white, cottony-looking forms. They were covered with brown spots. I later found out that these ball-shaped affairs were insect galls that were found only on oak trees. They were caused by a tiny insect called a gall wasp.

There is much mystery about how a gall is formed. Some say that the wasp stings the plant stem, then lays an egg. Others say it is caused by the bite of the wasp then the egg is laid. Whatever the case, a growth begins to form around the egg and eventually the gall is formed. Once the gall is complete the egg hatches into a larva, which gets its nutrition from the tissue inside the gall. When all the tissue is eaten the larva forms a pupa. When the pupa is mature, it splits open and an adult wasp emerges. It chews a small hole in the now dry gall wall and emerges. Then it flies off, ready to begin the life cycle all over again.

Certain species of oak galls were once used to make indelible ink.

To me, one of the strangest life cycles in nature is performed when an oak gall is formed.

Bud Jones

Chapter 11
Abbott's Sphinx Moth
(Sphecodina abbotti)

A lady once brought a caterpillar into my shop and asked me to tell her what it was. It was 3 inches long with a segmented body, covered with brown scale-like markings. It had an "eye" at one end. I was truly surprised and told her that I had never seen a caterpillar with an eye and I had no idea what it was. She let me keep it so that I might identify it, so after she left I got busy consulting my caterpillar books. I had to know what that critter was.

Later I identified it as the larva of a sphinx moth, called Abbotts Sphinx moth. The adult is brown colored dorsally on the front wings, while the hind winds are edged with light brown on the front side and dark brown on the back wing edge. When clinging to the side of a tree with its wings closed the moth is virtually invisible.

The larva goes through various stages, one of which forms the false eye. This convincing false eye has a white ring around it and it gives the appearance of a real eye because it is moist looking and seems to have a brown iris. That false eye is perhaps used to scare away potential predators.

The caterpillar feeds on wild grape leaves or Virginia creeper until it pupates, and eventually turns into an adult moth. Adults resemble hummingbirds, often hovering over flowers, seeking the nectar. They are also attracted to fermenting fruit.

The squirrel pictured has an unusually white crown. This is not normal.

Chapter 12
The Flying Squirrel
(Glaucomys volans)

The evening light was fast fading as I eased out of the woods on a day in October, after an afternoon of squirrel hunting. Suddenly, high in the tree tops I heard a rustling sound. I looked up just in time to see a flying squirrel emerge from an old woodpecker hole, climb quickly to the highest part of the tree, then leap out into space. It glided safely through the tree tops and then disappeared into the evening gloom.

This was my first introduction to this interesting mammal, which is common all over the United States. Due to its propensity for nocturnal living it is hardly ever observed by many people.

The flying squirrel is about six inches long from its nose tip to the base of its tail. It also has a soft, furry 4 inch tail which acts much like a rudder as it sails through the air. It is an olive brown color above and white below, with fairly long whiskers and big, dark, liquid looking eyes. It is well adapted to live the night life.

Perhaps the most interesting characteristic of this softly furred mammal is the flap of loose skin that attaches to the front legs, goes along the sides of the body, then attaches to the hind legs. When it leaps into space the legs are spread, forming a flat, broad surface, which allows the squirrel to easily glide through the air. In this way, gliding from a high perch, it can easily go the distance of a football field, controlling its movement while in the air by moving its legs and tail.

Flying squirrels live in old woodpecker holes, hollow trees, and similar places. Their den is lined with soft fur, feathers, etc., making a snug, warm nest. In April the female may have from 2 to 6 young. They eat insects, nuts, seeds, corn, and even baby birds and other meat that they may find.

Chapter 13
Jewelweed
(Impatiens capensis)

During the summer season many people pass by this unique plant, never noticing it or knowing of its usefulness. Normally it grows in damp places such as ditches, creek banks, wet spots in the ground and similar places. It can be a fairly tall plant, two or three feet perhaps, with a juicy, succulent, translucent stem. The small, pale orange flowers are splotched with reddish brown. Usually it blooms from mid-July until frost. The tiny seed pods, once formed, are very fragile, and when touched will literally explode, sending tiny seeds flying through the air. Another common name for jewelweed is touch-me-not.

Unknown to many people is that jewelweed, or touch-me-not, has the chemical make-up to cure poison ivy. If a person has been exposed to that troublesome plant, then by breaking a jewelweed stem and applying the sticky juice to the affected area, the itching and inflammation is very often cured.

Folks who know the curative powers of jewelweed over poison ivy often pull the whole plant, crush it in their hands, and boil it in a small boiler of water for about 15 minutes. Once the water has turned an amber color, it needs to cool. Then it can be applied with a clean rag or a paper towel to the affected area. I have seen some dramatic results with jewelweed while used in this manner. It can be stored in the refrigerator and used the whole summer long.

Bud Jones

Chapter 14
The Freshwater Jellyfish
(Phylum Cnidaria)

My knowledge of the freshwater jellyfish is very limited. So much so that, in fact, I never knew there was such a thing until the summer of 2010. A friend called me and asked if we had freshwater jellyfish in the Northwest Georgia area. I told him I had never heard of such a thing but I would check on it and find out. The next day he brought me a gallon jar full of water that contained about 8 of the critters. He said they came from a local lake. I was flabbergasted.

Upon examination I found the creatures to be opaque with 4 equally spaced black spots in the center. Those spots, I later learned, were the gonads, or sex organs. Around the rim of each jellyfish were the tentacles, string-like projections that dangled in the water. Each tentacle contains thousands of cells, and they are used to capture food – zooplankton – minute animal and plant life, and pass it into the mouth.

Drifting along in the water those tentacles are extended, the jellyfish waits for its prey to come in contact with its tentacles. Once this contact has been made the tentacles inject a poison into their prey, which paralyzes it, and then the food is transferred by the tentacles to the mouth of the jellyfish.

The life cycle of the freshwater jellyfish is very complicated; going through various stages, and would take a Marine Biologist to fully understand it. Suffice it to say that a fertilized egg eventually develops into a mature jellyfish.

The freshwater jellyfish is not harmful to humans.

Bud Jones

Chapter 15
The Buzzards Are Back
The Black Vulture (Coragyps atratus)
The Turkey Vulture (Cathartes aura)

Every year around November 15th, the city of Tallapoosa, Georgia, located 60 miles west of Atlanta and about 3 miles from the Alabama line, is inundated with buzzards, properly called vultures. Two species are included, the black vulture and the turkey vulture. They come from the northern states and join our resident buzzards to spend the winter in the area. It is believed that the big birds, which feed on dead animals, have a hard time finding food under the heavy snows of the North, so they come south in the fall where the pickings are easier.

During the day huge flights of the birds can be seen in the sky soaring, riding the hot air thermals that rise from the ground. Often they perch in trees around houses, causing messy driveways and yards. I personally have seen hundreds perched in an oak tree during the daylight hours. In the early morning when it is cold and the sun is shining, they often perch and spread their wings, absorbing the warmth of the sun before taking flight in search of food.

I know of one man who had a dead oak tree at the side of his driveway. The birds perching there made such a mess on his driveway that he finally cut down the tree. That solved his problem.

Winter passes into the blustery days of March. By the fifteenth of that month the visitors are gone, leaving only the resident buzzards to fill the void until the following November.

Bud Jones

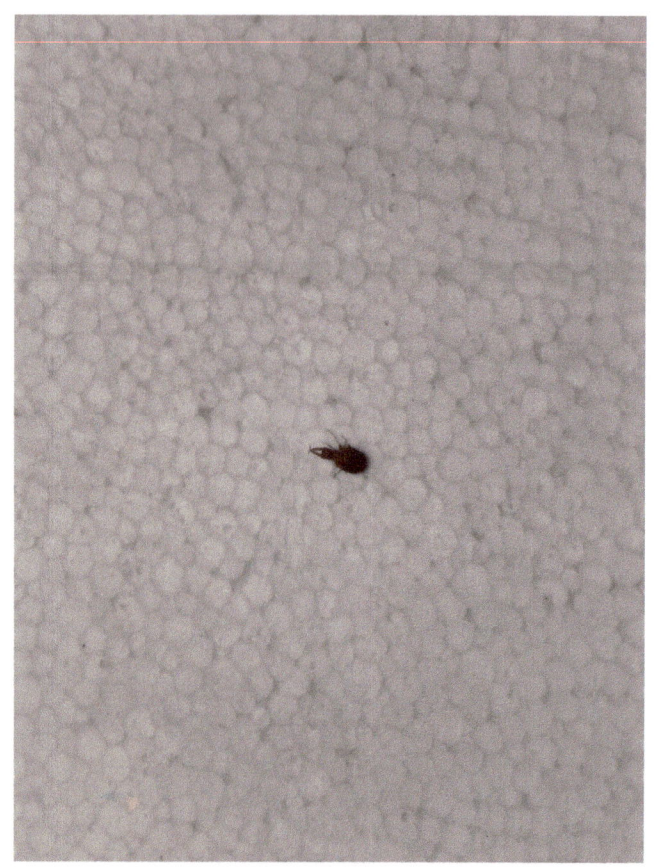

Chapter 16
The Ant Lion (Doodle Bug)
(Myrmeleon immaculatus)

"Doodle bug, doodle bug, come out of your hole."

These are words I used to say when, as a child, I would take a twig and stirred it around in an ant lion burrow. The common ant lion is truly a mysterious critter. The larva twirls around in dry dirt, usually under a shed or some other sheltered place, until it makes a conical hole in the dry ground. Then it covers itself at the bottom with sand and waits. Sooner or later an ant or some other insect comes by and falls in the conical hole. Because the sides are so loose and dry the hapless bug cannot climb out. Suddenly the larva darts out and sprays the bug with loose dirt, hastening its demise.

The ant lion larva seizes its prey with its large jaws and sucks out the body juices of the bug.

If the ant lion moves to a new location it moves to dry sandy soil. It can only walk backwards, similar to a crayfish, and as it does it leaves a distinctive trail in the dry dirt. When it comes time to pupate the ant lion larva burrows deeper into the soil and forms a sand covered cocoon. In this cocoon it gradually turns into an adult ant lion.

The adult ant lion resembles a dragon fly or a damsel fly. It has clear wings with a little brown spot at the end of each wing. One sure way to distinguish the adult from any other insect is the fact that each antenna is clubbed at the tip.

The adult ant lion is a very weak flyer. Their flight is erratic and clumsy, leaving one to wonder how they ever become airborne. The female, once she has lived out her few weeks of life, deposits her eggs on dry soil. The hatching larva burrows into the soil, makes a conical burrow, and the life cycle starts all over again.

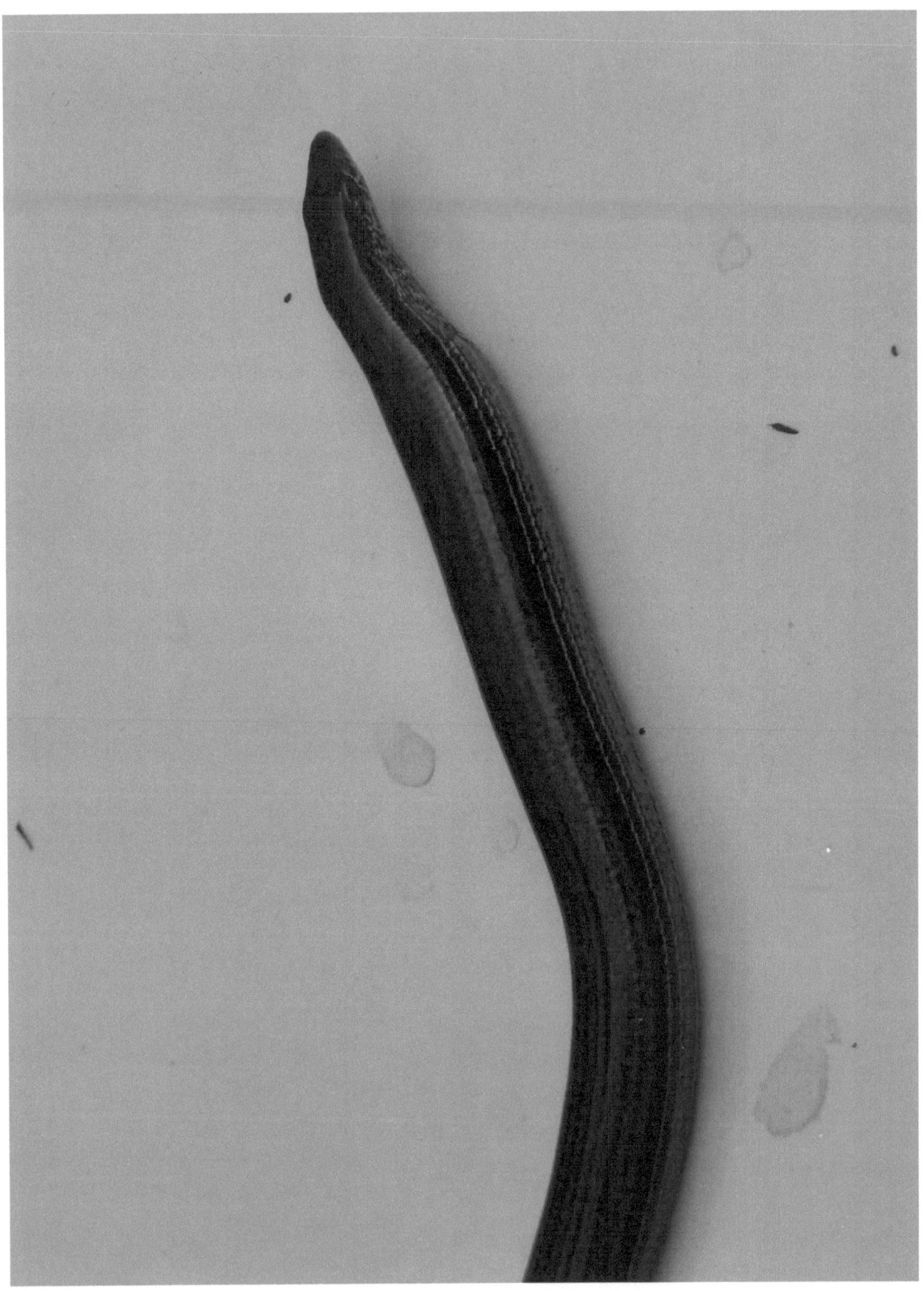

Chapter 17
The Glass Lizard
(Ophisaurus attenuatus)

One of the most unique animals in nature is the glass lizard, or joint snake, as it is often called. I have often watched one crawl away from what it perceived as danger, and I marveled at its jerky movements. For, despite what many people think, the glass lizard or glass snake is not a snake at all. It is a lizard without legs. On close inspection one can see ear openings and eye lids, something a snake does not have. This odd reptile also has small belly scales, while a snake has the wide belly scales that assist it in crawling.

A glass snake is about two-thirds tail. I once tried to catch a glass snake and as I grabbed it by the tail, it broke, leaving me holding a wiggling tail while the lizard got away. Like other lizard kin, the tail starts to grow back, but it never reaches its original length. Instead it often grows a stubby pointed tail that many people mistake for a stinger. There is a myth that the broken tail somehow comes back together on the body again—that is—the broken joint miraculously is reattached to the main part of the body. This has given rise to the erroneous name "joint snake."

Glass lizards eat other lizards, insects, snails, mice and young birds, small snakes, bird eggs, etc. The female lays eggs in a hole she scratches out with her nose and, once born, the young ones are on their own.

Bud Jones

Chapter 18
The Narrow-Mouthed Toad
(Gastrophryne carolinensis)

Most people are familiar with toads, which spend most of their lives on land and have rough skin. Frogs have smooth skin. I have included the narrow-mouthed toad because it has smooth skin and, since it is a burrowing toad, is seldom seen.

Usually about an inch long this toad can easily be identified by the transverse fold of skin that runs across the head, just behind the eyes. This peculiar amphibian has horny protuberances on its hind feet that are used for digging, and it is adept at disappearing rapidly into the litter of the forest floor when the occasion merits such behavior. Usually found in moist places the one pictured on the opposite page was trapped in a 5-gallon bucket that had a little water in it. When I released the poor fellow he was kind enough to pause for me long enough so that I could take his picture. These odd creatures mostly eat ants and termites but will also take most any insect that they can swallow. Like other of their toad kin they have to go to water to breed. The male attaches himself to the back of the female. Then he secretes a very sticky fluid that more or less glues him to the female, making him hard to dislodge by any other male frog. As the female lays her eggs the male fertilizes them. It takes a few days for the eggs to hatch into a tadpole. Then, in about two months, the little tadpoles become a full grown – though tiny – adult.

Bud Jones

Chapter 19
The Velvet Ant
(Dasymutilla occidentalis)

The velvet ant is not really an ant at all but actually belongs to the wasp family. The ones in our area (Northwest Georgia) are red and black. The female is wingless while the male sports a pair of blackish colored wings. Often called "cow" ant because folklore says that the sting of the female is powerful enough to kill livestock; and the sting, scientists say, is worse than the sting of any other North American insect. I know of a man who, knowing no better, picked up a female velvet ant. She gave him a terrible jolt (the male cannot sting) and he said it bothered him for 3 or 4 days.

The life cycle of these insects is very interesting. Females have a very hard body and do not seem to be bothered by the sting of another insect. When egg laying time comes they crawl into the burrow of other insects, such as wasps, bumblebees, and others. The female then lays her eggs on the larvae of these insects. When the velvet ant larvae hatches they devour their host, then they pupate, and later turn into an adult. To add insult to injury, pupation takes place in the burrow of their host.

Chapter 20
Lichen
(Thallophyta)

Lichens are around us nearly everywhere. We often walk on them, touch them, and see them daily yet most people are not aware of them. They are a remarkable group of plant life, growing on trees, rocks, and other places and are a good example of symbiosis in the world of nature.

A lichen plant is actually made up of two plants, an alga and a fungus. The alga, being a greenish color, manufactures food for the plant body through a process called photosynthesis. The fungus easily absorbs water and can hold the moisture for long periods of time. During dry periods it protects the alga and keeps it moist enough to produce food.

Most lichens are a gray-green color, the algae being green and the fungus white, thus giving the overall plants their gray-green color. Generally they form special reproductive bodies called soredia, or for the everyday layperson, they are called spores. Each of these bodies consists of a few algae cells surrounded by fungal cells. They form a powdery mass on the surface of the lichen and are carried away by the wind. Those spores alight on a host tree bark, rocks, logs, etc., and start another colony.

Some lichens, like "reindeer moss" furnish food for animals, while birds such as hummingbirds and blue-gray gnatcatchers use copious amounts to build their nests.

Bud Jones

Chapter 21
The Tent Caterpillar
(Malacosoma americanum)

As a boy I used to think that the huge masses of silk webbing that I often saw in cherry, apple, hawthorn and other tree species, was the work of spiders. It was many years later that I discovered that those big globs of webbing were actually the work of the webworm moth.

The moth itself is small, brown colored, with two white stripes going horizontal across the fore wings. In early spring the female moth lays a foamy ring of eggs around the twigs of trees. As soon as the spring foliage forms on the tree the larvae spin huge silken webs. It is a communal affair, often forming an "apartment" three or four feet high.

These larvae crawl out of this huge pile of webbing to feed on the leaves, leaving a shiny silken trail that they follow back and forth to the feeding area. At night or in bad weather they crawl back into the webbing for protection.

After a few weeks of use the webs become fouled with cast off skins and feces. Then the larvae fall to the ground. Once there they crawl under leaves and other debris and later spin a cocoon. When the adults hatch from the cocoon they go about the business of egg laying, thus starting the life cycle all over again.

Webbworm larvae have many enemies. Insects like assassin bugs suck out their body juices. Many species of birds, especially the yellow-billed cuckoo, feed on them heavily. During the summer the larvae can often be seen crawling on the ground as they seek a place to pupate. They are easily recognized, having a blackish-brown body with a white stripe going down the middle of the top side. On each side every individual segment has a bluish spot that encircles a tiny white speck.

Chapter 22
The Tiger Swallowtail
Two Butterflies In One.
(Papilio glacus)

I have been a butterfly collector most of my life. It has been more or less of a hobby-passion that I greatly enjoy. During all these years I have learned that Mother Nature has some odd tricks up her sleeve. For example, as a teenager I often caught the beautiful tiger swallowtail, but I was unaware that there was a black form to the species. In those days good field guides with colored pictures, etc., were nearly nonexistent, and if they had been available I could not have afforded one.

It was years later that I became a serious collector and developed an interest in correctly identifying my specimens. It was then that I learned that the beautiful blue-black swallowtails that I caught were really a variant, a phase or form, of the female tiger swallowtail.

Fast forward the years when my daughter Robin was going to college in North Carolina. One day she found a dead tiger swallowtail in the parking lot. The right wing was the normal yellow color, while the larger left wing was of the female blue-black color. I was flabbergasted the day she brought it to me and to this day I do not understand the mechanics of how this had happened. It must be rare, for in all my years of collecting I have never seen it. Certainly this is truly an oddity of nature.

Today, over 20 years later, that odd specimen remains one of my daughters prize possessions.

Bud Jones

Chapter 23
The Caddisfly Larva
(Trichoptera)

The caddisfly is moth-like yet it is not a moth, but belongs to the order of insects called Trichoptera. The larvae are aquatic and are most unusual in appearance.

Female caddisflies lay a gelationous string of eggs, usually on rocks or other objects close to or in the water. The eggs hatch, or turn into, larvae that faintly resemble caterpillars but are not as fleshy looking. Some have a pair of flat looking legs at the end of their body.

In most any clear creek or stream these larvae can be found, living under rocks and other debris, holding on to the bottom of the stream with their legs. They feed on the larvae of other insects that share their habitat.

When it comes time to pupate the larva encases itself in a silken cocoon – yes, it has silk glands, and it attaches itself to the surface of a rock. The front of the cocoon has a pair of mandibles that help it free itself so that it can swim to the surface later on. Actually the adult caddisfly is still in the pupae case but operates the mandibles and legs so that it can work its way to the surface of the water. Once free of the water, the adult emerges and goes about the business of being an adult caddisfly.

Certainly the life cycle of this peculiar insect is odd – even a miracle – that happens all around us, yet most people never know of its existence.

Chapter 24
The Amazing Swallowing Apparatus of Snakes
(Reptilia)

Despite years of persecution and downright hostility the group of animals belonging to the class Reptilia have managed to survive and in many places even prosper. Part of the reason for this special triumph is because of their ability to kill and swallow animals that are bigger than they are.

On the facing page in the top photo you see a king snake killing a larger red-bellied water snake. Note the tight constrictive coil of the king snake and the gasping-for-breath look of the water snake. Every time the hapless water snake sucked in air, the coil grew tighter, until finally death came.

In the bottom photo you see the water snake disappearing down the wide-spread mouth of the king snake, which has teeth that slant backwards, insuring that its prey will not escape. It is a slow process and, barring any problems, the king snake finally gets the job done. The question is, how can a snake swallow prey that is bigger than itself?

Human beings have jaw muscles that hold the lower jaw in place very firmly. There is no stretching of skin or muscle. On the other hand, the working apparatus of a snake jaw is quite complex. The upper and lower jaws are held loosely in place. When swallowing prey much bigger, the jaw muscles expand, the skin stretches, and the scales are separated. The body of the snake is carefully and slowly swallowed whole.

I once had a pet hog-nosed snake. I fed it toads, and one day it swallowed 15 toads. When I went to bed that night the hind legs of the last toad were still hanging out the mouth of the snake. Once a snake swallows its prey, powerful digestive juices slowly digest its body, and usually the snake will not need to eat again for a week or more.

Bud Jones

Chapter 25
The Mushroom That Grows In Manure
(Basidiomycete)

Many people eat mushrooms. Some folks buy them in the store while others pick them wild or raise their own. They are a popular part in the diet of people all over the world.

I do not profess to be a mycologist. In fact, mushrooms have never been an important part of my study. Still, I profess to have a fairly mild interest in them, mostly because of their color. Some are white, some red, yellow, gray, blue, and so on. I like to see them, especially in the fall, when their colors seem the brightest.

On the facing page you see a mushroom growing in cow dung. When I took this picture I wondered if it was merely happenstance or was the mushroom growing in the right place. I later learned that there is a group of these fungi that only grow in the dung of animals that have been grain fed. Some are edible, others are not. In fact, some are poisonous, and in some states it is illegal to possess them.

My point here is not to classify or describe mushrooms. It is simply to say that there is a mushroom that requires a special habitat. Like the monarch butterfly that must have the milkweed plant on which to lay its eggs; the scarlet snake that mainly eats reptile eggs; or the endangered snail kite that dines mostly on apple snails. There is a mushroom that requires a special place to live, and that is in a pile of manure.

Bud Jones

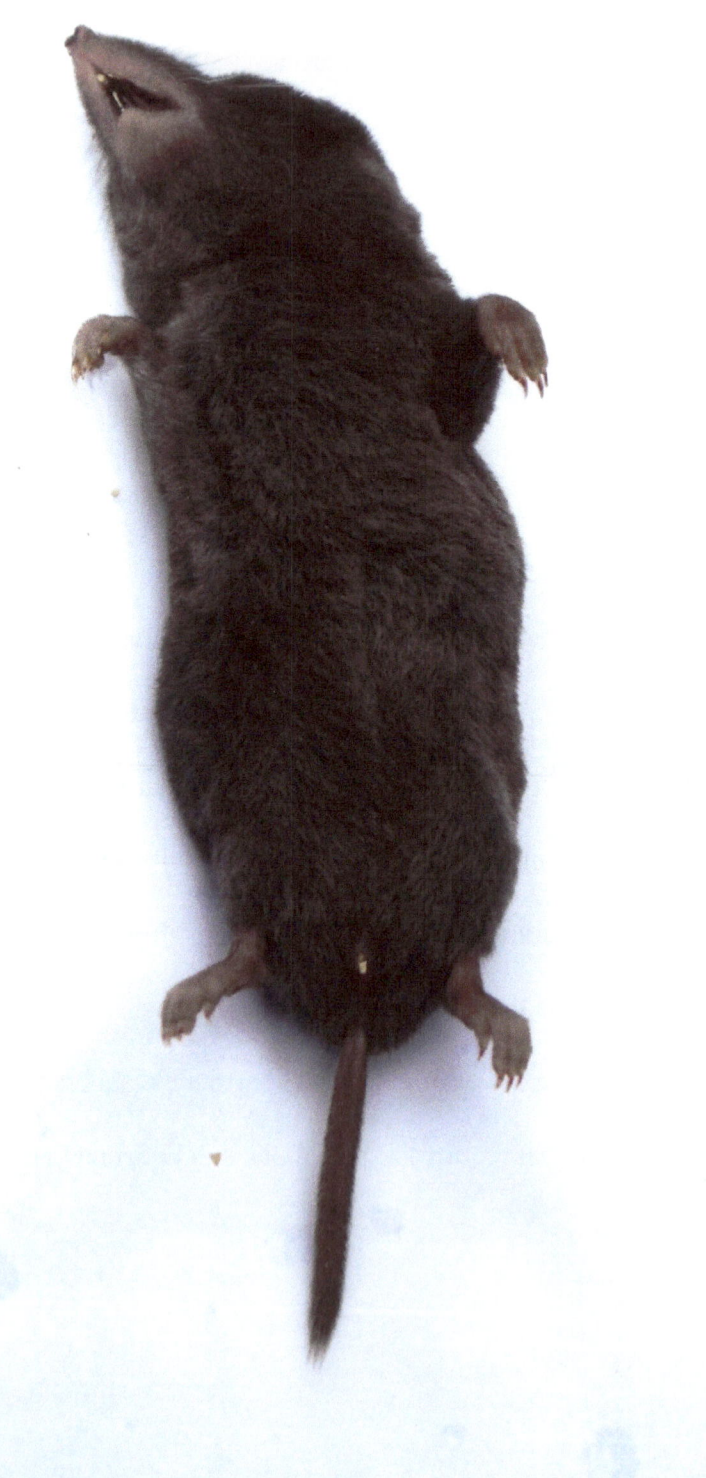

Chapter 26
The Short-Tailed Shrew
(Blarina brevicauda)

The short-tailed shrew is one of our most common mammals, yet it is hardly ever seen unless your housecat catches one and brings it to your door. Then it is most often mistaken for a mouse.

Shrews actually belong to the order Insectivora, which means that they mainly eat insects but will also dine on earthworms, centipedes, slugs, and snails. Actually they will eat most anything that they can overcome, such as mice, small reptiles, frogs, toads, even carrion. Some species have poison glands in their mouth. The poison, mixed with their saliva, helps to subdue their prey.

Short-tailed shrews are lead colored, with a 3 to 4-inch body and a 1-inch tail. They have no external ears, and their eyes are so small that they can hardly be seen. They live a solitary life, making burrows in the soft litter of the forest floor. Though their eyesight seems poor their long nose allows for keen smelling and their mass of whiskers for an accurate feel for their surroundings. They are very vicious animals and as such, we should be very glad that they get no larger than they are.

Their breeding season lasts all summer and 4 to 6 young are born to a litter. The burrow nest is often lined with hair from mice that have previously been killed. Though they are prolific animals they have a very short life span. Lucky are those who live over a year and a half.

Shrews are preyed upon by foxes, bobcats, owls and hawks, and other animals.

Bud Jones

Chapter 27
The Hercules Beetle
(Dynastes tityus)

A friend once brought me three large grubs. He had been digging in a wooded area around a big dead log when he decided to see what he could find in the lot itself. He was in for a big surprise, because he found three large grubs, white in color and three inches long, inside the dead wood. He brought them to me, and since I had never seen the like before I decided to try to raise them to maturity. I thought that they might be the larvae of the eastern hercules beetle, but I was not sure. I placed them in a fairly large container and covered them with a layer of rotten wood. Then I waited. Every week or so I dampened the wood and kept watch.

Often I became doubtful because I feared, after waiting for weeks and weeks that they had died. Every so often I would carefully dig down in the wood, only to find the grubs plump and healthy looking. Sometimes I even forgot about them. Then one day I opened the plastic lid and, lo and behold, a big fat, greenish, spotted beetle appeared on the surface. It had been nearly two years since I put them in that plastic container.

I had guessed right. It was an adult eastern hercules beetle, a female, for she had no horns. Only the male has an upper and lower horn. This species is the largest and heaviest beetle in North America. The larvae, as I found out, take two years to mature, and while they feed on decaying wood, the adults are attracted to fermenting sap and fruit.

Two grubs hatched. The third never did. It was truly an experience for me to see the total transformation of a larva that was bigger than the adult, yet it turned into a giant shiny beetle.

Surely He who created everything wanted to give us something to marvel about, and the life cycle of the eastern hercules beetle gives us much to ponder.

Chapter 28
The Chuck-Will's-Widow
(Caprimulgus carolinensis)

The chuck-will's-widow was named after its song – "chuck-will's-widow, chuck-will's widow, chuck-will's-widow" a long drawn out series of repetitive notes that seem to go on forever as they are carried on the breeze of a hot summer night.

This bird belongs to the goatsucker family because colonists thought that they actually "milked" their goats. Many people have heard the call of this bird of the night, yet have never seen the caller. Its song is often confused with that of the whip-poor-will, but a careful observer can easily tell them apart.

The chuck-will's-widow is about 12 inches long and of a mottled brown color. Its feet are so small and weak that it does not perch crossways on a limb but parallel to it, so that its whole body might rest on the limb.

However, even the small feet are not what makes the bird so odd. What sets it apart is its big mouth, so big that it has a 2-inch gape, really big for a 12-inch bird. This big mouth is surrounded by whiskers or bristles, so that as the bird flies through the night sky those bristles help to guide an insect into the wide open mouth, for insects are the main food of this avian flytrap.

On the facing page you can see the opened mouth of a chuck-will's-widow. It had been hit by a car and found by the author.

For some strange reason the chuck-will's-widow is disappearing from its usual haunts and scientists are not sure why. Perhaps insecticides could be the reason, or maybe fragmented forest land, or perhaps both. Whatever the cause it would be a shame to lose such an unusual bird.

Bud Jones

Chapter 29
What About The Sex Life Of Snakes. How Do They Breed?
(Reptilia)

Many of us have observed various mammals and birds breeding, such as dogs, cats, horses, chickens, etc., but not many people have been privy to the breeding cycle of snakes.

Luckily a friend called me one day to tell me that he had taken a picture of two gray rat snakes that were actually in the act of copulation. He kindly gave me a photo which appears on the opposite page. He said that he actually picked up the two reptiles while they were busily engaged and took them out of his yard. In all that time they never separated but stayed entwined.

Normally, when a female snake is ready to breed, she releases a chemical substance called pheromones. When a male snake is encountered she raises her tail as a sign that she is willing to copulate. The male comes alongside her and they wrap around, or entwine, with each other. The male inserts his forked penis, called a hemi-penis, into her cloaca, and mating takes place. Often the two snakes rub against each other for a while before the actual mating takes place. During this process of clinging together they can be picked up and moved around, but once the act is completed they part and go their separate ways, probably never to meet again.

Bud Jones

Chapter 30
The Praying Mantis
(Tenodera aridifolia)

Once, years ago, I was standing in my daughter's yard admiring the butterflies that swarmed around her butterfly bush. Soon I noticed an accumulation of butterfly wings under the bush. I was astounded. There were monarch wings, swallowtail wings, fritillary wings—on and on. I called my daughter Cherry and we both stood there, looking perplexed at the accumulation of wings under the bush. Suddenly Cherry pointed toward the middle of the bush.

"Look," she said.

Then I saw a live gulf fritillary butterfly held in the vise-like grip of a praying mantis. I shook the limb they were on and, to my amazement the mantis released the butterfly and flew away. Then I had an answer as to where all the butterfly wings came from.

The praying mantis, like all insects, has six legs. At rest the two large front legs are folded under the body, giving the appearance of a praying posture. The mantis is camouflaged well and it lies quietly in wait until some hapless insect—in this case, a butterfly—comes in range. Then those two strong front legs grab the prey, tear it apart, and eat it alive. When food becomes scarce in one place the mantis flies to another location.

Female mantis lay their eggs in rows, making a gray mass that looks like foam, usually on a twig. This odd looking mass becomes hard, and when the eggs hatch they are tiny replicas of the adult.

The bug-eyed appearance of the praying mantis and its posture of praying surely make it one of our most odd insects.

Chapter 31
Pinesap – A Strange Plant
(Monotropa hypopithys)

Pinesap is a plant that has no chlorophyl. It is a member of the Indian pipe family and has a soft, hairy stem. Normally it grows in clusters around pine and oak trees.

Since pinesap has no chlorophyll it cannot manufacture its own food. To cover this deficiency the roots of the plant are covered in tiny fungi. These fungi enable the plant to absorb nutrients from the soil, thus enabling it to function as does a normal plant.

The pinesap itself has over-lapping scales on the stem which serve as leaves, while the flowers are arranged only on one side of the stem. The yellowish-red color darkens with age. When the tiny flowers mature a small ovoid capsule is formed to start another generation.

Pinesap is edible either raw or cooked and contains essential oils. It is not a common plant, yet what makes it so different is that, unlike a normal flower-bearing plant which contains chlorophyl pinesap cannot do this. Yet it has adapted its own method of survival, which certainly makes it quite different from other plant species.

Bud Jones

Chapter 32
The Flat-headed Worm
(Land Planaria)

Once a man called me all excited.

"Bud", he said, "Can you come out to my shop? I've got something on the side of my building. I can't tell what it is."

Always the curious one when it comes to something natural, I was at his place of business in less than 10 minutes. He led me around to the north side of the building and pointed.

"Now what in the world is that?"

Clinging to the side of the building was a flat-headed worm, sometimes called a shovel-headed worm. I told my friend what it was and as I stooped to take the 6-inch critter off the building it quickly stretched to a length of about 15 inches. My friend was flabbergasted.

Flat-headed worms are rarely seen. Sometimes after a heavy rain they will come to the surface of the ground, yet then they are usually mistaken for the common earthworm. They can also be found under rocks, rotting logs, etc. Their semi-circular heads are flat and reminiscent of a shovel or a spatula in appearance. They do not have eyes but can perceive light. When they move about they leave a slime trail, just like a snail or slug.

Strangely enough they can lay eggs, but can also break off a segment of tail, which in a few days can grow into another worm. Their food is earthworms, which they wrap around and use their mouth (which is located about mid-body) to suck out the body juices. Oddly enough that same mouth also serves as an anus.

Now—can you think of anything more odd that a flat-headed worm?

Bud Jones

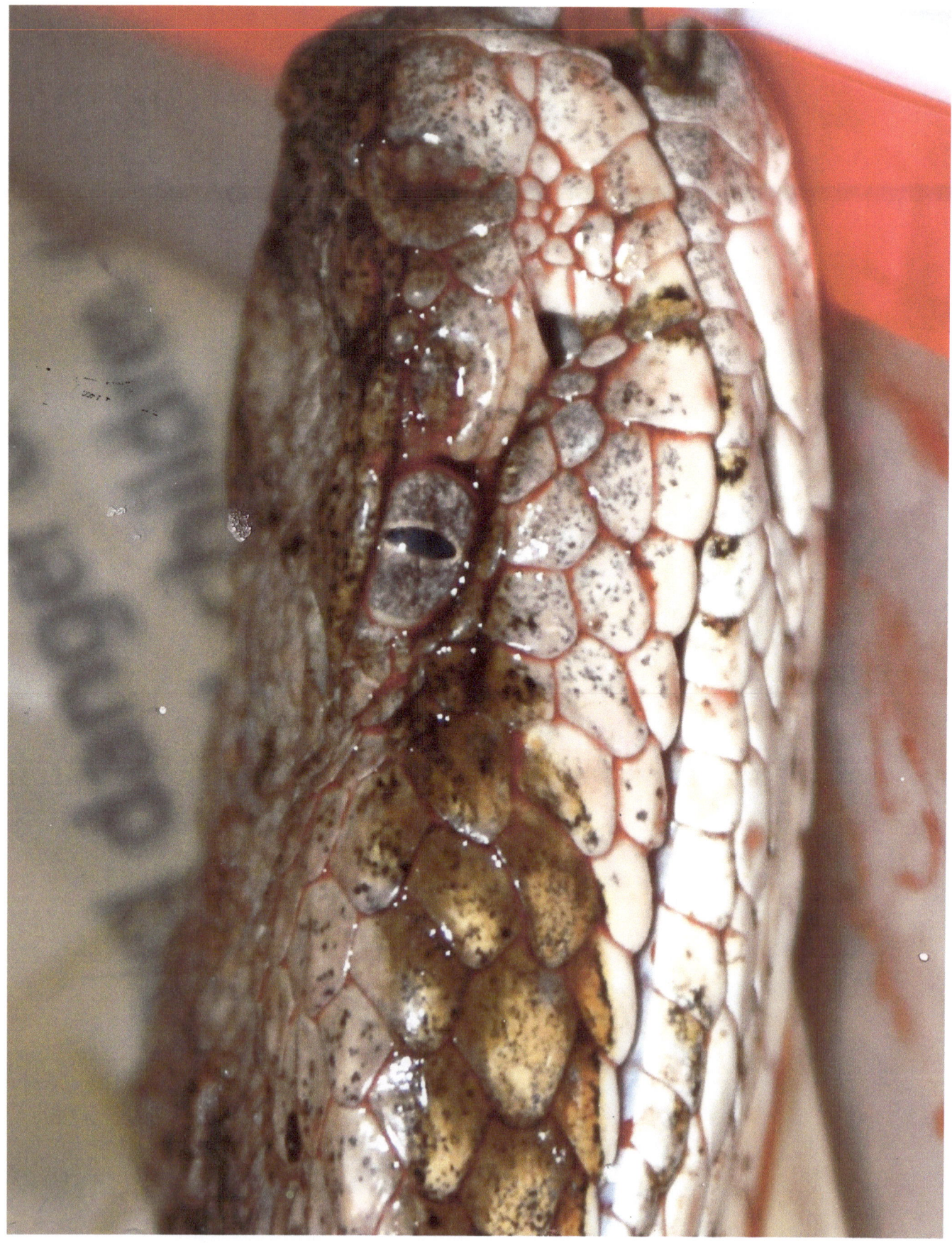

Chapter 33
The Guidance System of the Canebrake Rattlesnake
(Crotalus horridus)

Many people have a misunderstanding of how snakes procure their food. For example, black racers, king snakes, coach whips and other similar snakes are fairly slender and can move about with alacrity. They can grab their prey, wrap around it, and squeeze it to death. On the other hand, poisonous snakes, such as rattlesnakes, copperheads, water moccasins, etc., are bulky and not as agile as the first mentioned group. Therefore, over eons of time they have developed a more sophisticated method of capturing prey. We will use the canebrake rattlesnake as an example.

The canebrake rattlesnake is called a pit-viper. Halfway between the eye and nostril is a small opening called the pit. This opening is a heat sensor, and by using this pit the snake can detect the presence of a warm blooded animal. Lying in complete darkness the snake can "feel" an approaching rat or rabbit, for example. In this way it is guided to strike just at the right moment. If the prey, such as a rabbit, runs off before it dies, the rattlesnake using its heat sensor is able to trail its victim. Then it swallows it whole.

Another feature of pit vipers is that their fangs are hollow. When biting their prey the poison is squeezed by muscles surrounding the poison gland and forced through the fangs, which have a tiny hole at the end of each fang. When not in use these fangs are folded back into the roof of the mouth. If one fang is broken, it is replaced by a fang behind it, thus assuring the vipers of always having a way to secure their prey.

Bud Jones

Chapter 34
The Cicada – A Life Spent Underground
(Family – Cicadidae)

Many nights I have spent on the banks of the Tallapoosa River in my sleeping bag, being lulled to sleep by the see-sawing "song" of cicadas. In my area of the country they are called July flies, but no matter what the name they are truly an odd member of the insect world.

Female cicadas lay their eggs on twigs of living trees. When the eggs hatch the so-called nymphs fall to the earth, then dig their way into the soil. Once underground they attach themselves to plant roots and suck the sap from them. They remain underground for several years, usually from two to five. One species, the seventeen-year cicada (often called the seventeen-year locust) only emerge every 17 years.

When the time comes to free itself from its nymph shell, the nymph comes out of the earth and crawls up a tree or similar object and attaches itself with its strong legs. The shell splits down the back and an adult cicada emerges, leaving its ghostly looking shell still clinging to its original perch. Once its body juices flood the whole body and the outer exoskeleton hardens, the adult cicada is ready to begin its own life.

Cicadas have a body perhaps a little over two inches. Their wings are clear and they have a robust build. Once their wings are dry they fly away, only to begin again a life that is lived mostly underground. The facing page shows the exoskeleton left behind by a newly hatched adult.

Chapter 35
Regeneration – The Five-Lined Skink
(Eumeces inexpectatus)

Wouldn't it be wonderful if we, as humans, lost an arm or leg and could grow another body part to replace the lost one. We know that such a feat is impossible, yet in some animals it can happen. The photo on the left is a good example. It shows a member of the lizard family, a five-lined skink, that is regenerating a new tail. Normally the tail of this species is about one and a half times as long as the body length. However, as the photo clearly shows, the original tail has been lost and a new one is growing back to replace it.

A lizard has a special adaptation. If a predator catches it by the tail, that particular part of the anatomy breaks easily, leaving only a wiggling tail for the predator to look at. The lizard makes its getaway. Soon a new tail begins to grow and in time another one is formed.

Crayfish are also a good example of regeneration. If one of their "pinchers" is accidently lost, a new one will grow back to take its place. A starfish can lose a leg and it will be replaced.

So, if you see a lizard with a stumpy tail, simply keep in mind that it is Mother Nature's way of survival for some species.

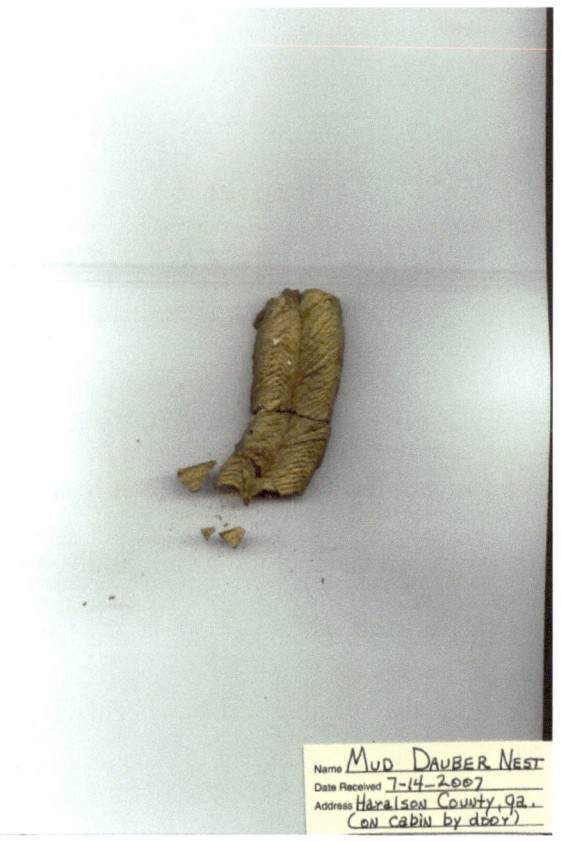

Name: Mud Dauber Nest
Date Received: 7-14-2007
Address: Haralson County, Ga. (on cabin by door)

Name: Contents of Dauber Nest
Date Received: 15 spiders, 1 pupa
Address: and egg cases. Also shows how end is sealed

Chapter 36
The Mud Dauber
(Trypoxylon politum)

A most peculiar example of an insect that has a special adaptation is the "dirt dobber", as I used to call it, or more correctly the mud dauber. This member of the wasp family is well known for gathering bits of mud and making a cylindrical nest, attaching it in a sheltered place to a wall, a piece of furniture, under porches and car ports, or in some similar location. Once the nest is made the mud dauber goes in search of spiders.

I have seen a mud dauber walking energetically through grass and weeds as it searched for prey. Once located, the hapless spider is stung, leaving it paralyzed but not dead. Then it is taken and deposited in the mud nest. When enough spiders are collected the dauber lays an egg in the mud nest, then that nest is sealed off with mud. When the egg hatches the larva dines on the live, but paralyzed spiders. When its food source is all gone the larva pupates and later turns into an adult. Soon it goes about the business of being a spider killing mud dauber.

The survival of the larva of mud daubers on paralyzed spiders is certainly a unique way that the continuance of a species is carried on.

Chapter 37
The Armadillo
(Dasypus novemcinctus)

If God had fun creating the mammals He must have clapped His hands when He made the armadillo. It is indeed a strange animal. With a hard, scaly skin, it is about the size of a small dog, with nine movable bands in the middle of the body that allow it to move freely. Normally it is a gray-brown color but is often colored like the dirt in which it has been digging. Its armor plated, segmented tail is nearly as long as its body and is sharp pointed. The feet have long claws that are well adapted for digging.

Armadillos eat grubs, insects, etc. that they dig out of the ground. Their propensity for digging very often gets them on the bad side of home owners, for they can quickly mess up a pretty lawn, leaving softball-sized holes where they dug for food.

These mammals have some strange traits. One is that they have no teeth, so that is the reason for their soft diet. Another trait is their manner of birth. Female armadillos give birth to four young. They are all of the same sex, either all males or all females. There are no mixed broods.

Most authorities say that armadillos cannot stand cold weather. However, in my part of Northern Georgia it can get in the low teens often, yet they seem to thrive here. They probably stay in their burrows if the weather is too severe. I have seen them in the daytime, but normally they are nocturnal.

Chapter 38
The Ichneumon Wasp
(Megarhyssa atrata)

The ichneumon wasp is a very strange creature indeed. The female has three hair-like projections that hang from the tip of her abdomen. She searches through the woodlands for a dead hardwood log that is infested with the larvae of the pigeon tremex wasp. Once located, the wasp uses her antennae to detect the scent of a fungus that is in some way associated with the pigeon tremex larvae.

Using the hair-like projection closest to the middle of her body, called an ovipositor, she drills a hole maybe 6 inches deep, down to the pigeon tremex larva. The ovipositor itself secretes a chemical that softens the wood fibers of the log. Then an egg is laid, going through the long ovipositor, onto the pigeon tremex larva. When the egg hatches the larva feeds on its host. The other two hair-like projections are used as props when the female ichneumon wasp is drilling into the wood. Each egg is formed into a long strand so that it can easily pass through the long ovipositor.

Once the host larva is devoured the ichneumon larva spins a cocoon and eventually turns into an adult ichneumon wasp, which emerges in the spring and starts the life cycle all over again.

Bud Jones

Chapter 39
The Braconid Wasp
(Cotesia congregate)

There is a tiny wasp, less than half an inch long, called a braconid wasp. This small insect lays its eggs on the larvae of butterflies and moths. The photo on the opposite page shows a tomato hornworm that has been parasitized by one of these wasps.

The female wasp lays her eggs on the caterpillar. When the eggs hatch the larva bores its way into the body of the hornworm. Slowly the larva, in this case more than one, slowly devours the body of the hornworm from the inside.

Forty or fifty larvae gradually eat the insides of the caterpillar, yet it is still able to function. When the larvae are mature, each one tunnels to the outside of the hornworm and forms a cocoon. The photo on the opposite page shows more than 70 cocoons formed on the body of the hornworm, each resembling a grain of rice.

When mature, each cocoon splits open and an adult braconid wasp flies off to greet the world, while at the same time they are looking for a caterpillar so that they might carry on their own life cycle.

The tomato hornworm, its body decimated from the work of the many larvae, dies, never to complete its life cycle.

Bud Jones

Chapter 40
Another Species of Ant Lion
(Glenurus gratus)

Most of us, when we hear or read the words ant lion, think of the familiar doodle bugs that build their little conical traps under a shed or some other sheltered place. However, there is another species of ant lion that has similar habits.

The ant lion pictured on the opposite page lives in natural holes in trees. The female lays her egg in the sawdust of a tree hole. When the larva hatches it makes its own trap in the sawdust, and then hides at the bottom of the trap. There it patiently waits for an ant or some other insect to come along and fall into the trap. It then is devoured by the ant lion larva.

When the Master created the world He formed some odd creatures. Surely this ant lion is one of the oddest.

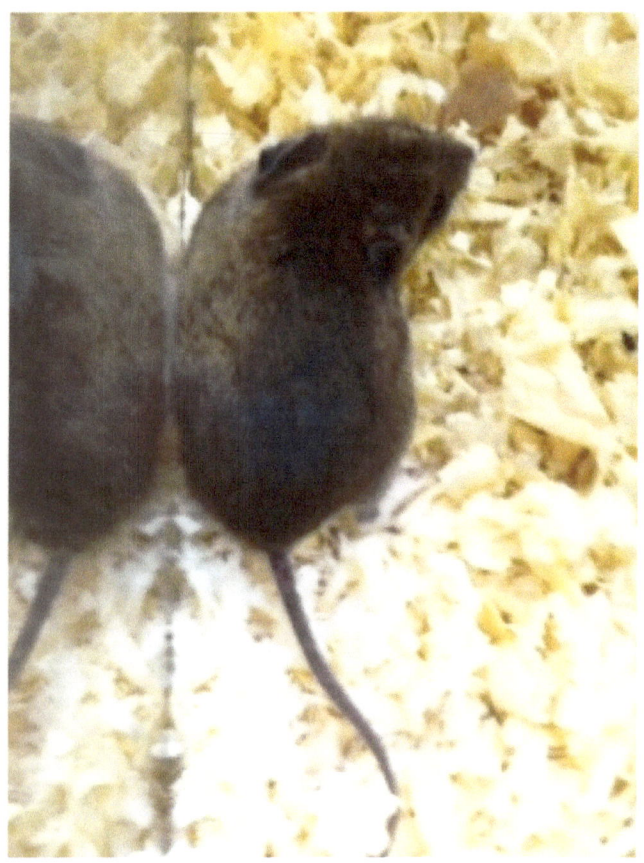

Chapter 41
The Vole
(Microtus)

I had read about voles and knew that we had them in our area, yet I had never seen one. Then one day a lady brought a live one to me and I instantly knew what it was.

Voles resemble mice, and many of us have probably seen one scoot across the road or path, and thought we had seen a mouse, for their resemblance is very similar. Their body is usually 4 to 5 inches long with a stubby tail. They make burrows in grass and leaf litter and are hard to see in such a habitat. Often they might be seen as they race from one burrow to another. They are active both day and night.

Voles eat roots, grass, fruits and seeds in the fall, and bark in the winter. They live in small family groups and a female may have up to four litters a year, consisting of 1 to 5 in a litter.

In the South they breed the year-round.

Sometimes voles may raid crops or girdle tree seedlings, but this damage is negligible. They fall prey to owls, hawks, foxes, bobcats, and other animals and in some areas the vole is an important item in the food chain.

www.ingramcontent.com/pod-product-compliance
Lightning Source LLC
Chambersburg PA
CBHW051022180526
45172CB00002B/437